ANAS EL AZIZI EL ALAOUI

Urban transformation in the age of globalization:

ANAS EL AZIZI EL ALAOUI

Urban transformation in the age of globalization:

Towards a deeper understanding of the spatial duality between ancient and contemporary cities

ScienciaScripts

Cover image: www.ingimage.com

This book is a translation from the original published under ISBN 978-620-6-71614-3.

Publisher:
Sciencia Scripts
is a trademark of
Dodo Books Indian Ocean Ltd. and OmniScriptum S.R.L publishing group

120 High Road, East Finchley, London, N2 9ED, United Kingdom
Str. Armeneasca 28/1, office 1, Chisinau MD-2012, Republic of Moldova, Europe
Printed at: see last page
ISBN: 978-620-7-84816-4

URBAN TRANSFORMATIONS IN A GLOBALISED WORLD: TOWARDS A DEEPER UNDERSTANDING OF THE SPATIAL DUALITY BETWEEN ANCIENT AND CONTEMPORARY CITIES

ANAS EL AZIZI EL ALAOUI [1]

SUMMARY

In the context of urban globalisation, cities face major challenges in managing the spatial duality between the old and the contemporary city. On the one hand, the old city is an integral part of urban identity and heritage, embodying the traditional urban fabric and heritage architecture that reflect the values and identity symbols of society. However, this ancient city is facing major challenges such as poverty, unemployment and social and economic exclusion, threatening the erosion of its identity and urban heritage. On the other hand, the contemporary city is a showcase for modernity and progress, characterised by modern buildings and state-of-the-art infrastructure, embodying aspirations for innovation and economic development. Nevertheless, the contemporary city is bringing about radical changes in lifestyles and societal values, which can give rise to social and cultural tensions. In addition, tensions and contradictions are emerging between the two cities over the use of land and resources, exacerbating social and economic divides and threatening the cohesion of society.

Keywords: Urban globalisation, spatial duality, ancient city, contemporary city, urban identity, urban heritage, marginalisation, economic development, social cohesion, social and cultural tensions.

INTRODUCTION

Globalisation has brought about profound urban transformations, significantly shaping the landscape of cities around the world. This complex dynamic has created a clear spatial duality between ancient cities, with their rich historical heritage, and contemporary cities, emblematic of modernity. As Béal, Rousseau and Vignon (2018, p. 11) point out, "globalisation is producing an acceleration in the pace of urban transformation, resulting in a growing divide between inherited and contemporary spaces". On the one hand, ancient cities, steeped in a long history, embody a distinctive identity and a valuable cultural heritage. However, as Djament-Tran and Fagnoni (2015, p. 8) note, "globalisation is exerting increasing pressure on these heritage spaces, calling into question their authenticity and uniqueness".

On the other hand, contemporary cities, characterised by bold architecture and state-of-the-art infrastructure, symbolise progress and innovation, but sometimes struggle to fit harmoniously into the existing urban fabric (Gravari-Barbas & Veschambre, 2005).This spatial duality raises a number of questions: How can we reconcile the preservation of our historic heritage with the imperative of contemporary urban development? What are the strategies for integrating these two facets in a sustainable way? How can we ensure social and territorial cohesion within these contrasting urban areas? How can we guarantee social and territorial cohesion within these contrasting urban areas? Faced with these challenges, it is vital to deepen our understanding of the multi-dimensional dynamics at work, in order to develop integrated approaches and appropriate urban policies capable of harmoniously enhancing historical heritage while embracing the opportunities offered by modernity. In-depth reflection on this spatial duality is essential if we are to shape cities that are sustainable, inclusive and respectful of their identity, in the changing context of urban globalisation.

1. THE GLOBALIZATION: MULTIDIMENSIONAL DYNAMICS SHAPING THE GLOBAL URBAN

Landscape globalisation globalization is characterised by transnational flows of ideas, capital, goods, services and people that transcend geographical borders. This multidimensional is shaped by economic such as trade liberalisation and the rise of multinational companies, but also by socio-cultural factors such as the spread of lifestyles and consumption patterns. Technological advances, particularly in transport and communications, have also played a key role in facilitating the global movement of people, information and capital. Under the combined effect of these dynamics, cities have become strategic hubs of globalisation, concentrating economic activities of activities,the command infrastructures and migratory flows, while at the same time having to contend withnew socio-spatial disparities.

1.1 Global definition of urban globalisation: the reciprocal flow of ideas, practices and activities across borders

Urban globalisation refers to a multi-dimensional process in which cities have become key nodes in a worldwide network of exchanges and interactions. This phenomenon involves an intense and reciprocal circulation of ideas, practices and economic, social and cultural activities across national borders (Saskia Sassen, 2001, p. 37). It is a complex dynamic that transcends traditional geographical boundaries, fostering a growing interconnection between urban spaces on a global scale. According to Marie-Christine Fourny (2008, p. 12), urban globalisation is characterised by "multiform flows of information, capital, goods, services and people on a planetary scale, linking cities together in a vast global network". This definition highlights the multidimensional nature of the process, encompassing not only economic flows but also social, cultural and human flows. For Dominique Lorrain (2011, p. 23), urban globalisation is "a process by which cities are integrated into global networks for the exchange of goods, services, information, capital and people". This author underlines the importance of the global networks in which cities are interwoven, enabling the intense, multidirectional circulation of various resources and activities.

Figure (01) : Globalisation in operation and in debate

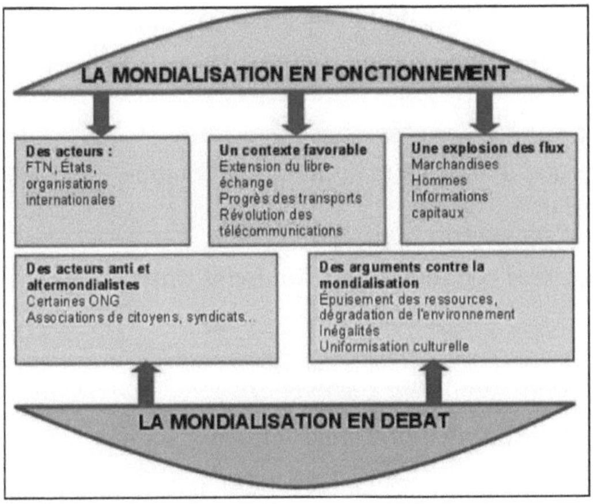

Source : http://thehumbertaker.wifeo.com/chapitre-2-geographie-ts.php

From this perspective, urban globalisation can be seen as a phenomenon in which spatial and temporal boundaries become blurred, allowing an interpenetration of cultures, lifestyles and practices across the world (Cynthia Ghorra-Gobin, 2015, p. 8). Cities are becoming cosmopolitan crossroads where local and global influences meet and mingle, giving rise to unique and constantly evolving urban dynamics.

1.2 Mechanisms of urban globalisation: multidimensional driving forces

Urban globalisation is a process fuelled by a constellation of multidimensional driving forces, which are profoundly shaping the dynamics and landscape of cities around the world. Key drivers include transcontinental economic and financial flows, driven by trade liberalisation, market deregulation and the rise of multinational companies.

The revolution in information and communication technologies has also played a decisive role, facilitating the instant circulation of data, capital and ideas, and redefining the relationship between space and time in the world's metropolises.

At the same time, international migration and increased human mobility have led to a demographic reconfiguration of urban areas, confronting cities with new challenges of social and cultural integration. The combined forces of economic, technological and human globalisation have profoundly reshaped the landscapes, flows and dynamics of contemporary cities.

1.2.1 Transcontinental economic and financial flows: their impact on urban structure

Transcontinental economic and financial flows are a fundamental aspect of urban globalisation, exerting a profound influence on the structure and dynamics of cities around the world. These flows, facilitated by the liberalisation of trade and the growing integration of markets, have led to a reconfiguration of urban landscapes, both physically and functionally.According to Cynthia Ghorra-Gobin (2015, p. 42), "cities have become strategic nodes in global networks of production, trade and finance, attracting high-value investment and economic activityadded value'. This concentration of capital flows and economic activities has led to a restructuring of urban areas, with the emergence of business districts, commercial zones and cutting-edge financial centres. This dynamic is reinforced by the increased mobility of multinational companies, which seek to optimise their presence in strategic cities. As Saskia Sassen (2001, p. 78) points out, "multinational companies have played a key role in the restructuring of cities, relocating their activities and creating new centres of economic attraction". This process has also led to a physical transformation of cities, with the construction of skyscrapers, housing complexes and modern transport infrastructure to meet the needs of companies and investors (Marie-Christine Fourny, 2008, p. 37). These developments have often been accompanied by processes of gentrification and population displacement, highlighting the social and spatial tensions inherent in urban globalisation. At the same time, financial flows have encouraged the emergence of global financial centres, such as London, New York and Hong Kong, which concentrate a large proportion of financial services and capital management activities (Dominique Lorrain, 2011, p. 112). These cities have become strategic hubs for international investment, attracting skilled labour and state-of-the-art infrastructure. However, these urban transformations driven by economic and financial flows are not without their social and environmental impacts. As Jacques Lévy (2008, p. 68) points out, "economic globalisation has led to increased spatial and social fragmentation in cities, with increasing polarisation between affluent and deprived areas".

Map (01) : Poles and flows of globalisation

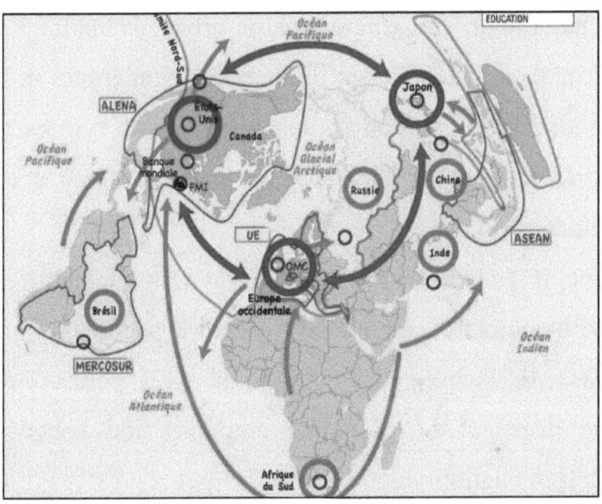

Source :https://h-g.jimdofree.com/ts-g%C3%A9o/th%C3%A8me-2-les-dynamiques-de-la-mondialisation/la-mondialisation-acteurs-flux- d%C3%A9bats/

1.2.2 The revolution in information and communication technologies: Reshaping urban spaces

The revolution in information and communication technologies (ICTs) has profoundly transformed urban dynamics, redefining concepts of space and time in cities. This digital revolution has encouraged the emergence of new forms of spatial organisation, challenging traditional models of urban planning and development.According to Manuel Castells (1998, p. 379), "the new information technologies are at the root of the formation of spaces of flows that are gradually replacing the space of places". This notion of flow spaces refers to digital networks that transcend physical borders and enable the rapid circulation of information, capital and people.In this context, cities have become key nodes in these digital networks, concentrating ICT infrastructure and services. As Cynthia Ghorra-Gobin (2015, p. 92) points out, "cities have been transformed into interconnected technological platforms, offering an environment conducive to the development of activities linked to the digital economy".

This dynamic has led to the emergence of urban technology hubs, such as Silicon Valley in the United States or the Cité Numérique in Brest, France. These areas concentrate high-tech companies, research centres and innovative start-ups, creating ecosystems conducive to innovation and economic growth (Pierre Veltz, 2000, p. 147).

At the same time, ICTs have enabled new forms of work and lifestyles, such as teleworking and e-commerce, which have reshaped patterns of mobility and use of urban spaces. As Jacques Lévy (2008, p. 112) points out, "ICTs have encouraged the dispersal of economic activities and called into question traditional models of centrality".

However, this digital revolution also raises issues in terms of equitable access to ICT infrastructure and services. As Marie-Christine Fourny and Marie-Flore Mattei (2010, p. 67) point out, "the digital divide is a reality in many cities, where some neighbourhoods are disconnected from digital networks, reinforcing social and spatial inequalities". Thus, the reshaping of urban spaces by ICTs raises challenges in terms of planning and development, requiring an integrated and sustainable approach to ensure equitable access to technologies and a balanced distribution of the benefits of the digital revolution.

1.2.3 Migration and global human mobility: challenges and opportunities for cities

Migration flows and human mobility on a global scale have become an essential component of urban globalisation, profoundly shaping the social, cultural and economic landscapes of cities. This multi-dimensional phenomenon raises both challenges and opportunities for urban areas.

According to Cynthia Ghorra-Gobin (2015, p. 125), "cities have become hubs of international mobility, attracting economic migrants, refugees and highly skilled workers". This diversity of migration profiles has led to a reconfiguration of demographic and social dynamics within cities. On the one hand, low-skilled labour migration has often been associated with problems of integration, insecure housing and spatial segregation. As Catherine Withol de Wenden (2013, p. 87) points out, "the disadvantaged neighbourhoods of large cities are often home to immigrant populations facing difficulties in accessing housing, employment and public services".

However, the migration of highly qualified workers, attracted by the economic opportunities offered by cities have also helped to shape urban landscapes. Saskia Sassen (2001, p. 192) points out that "global cities attract a skilled and cosmopolitan workforce, creating multicultural spaces and fostering innovation and creativity". This cultural diversity has also had an impact on urban landscapes, with the emergence of ethnic neighbourhoods, shops and places of worship reflecting the richness of migrant identities. As Emmanuel Ma Mung (2009, p. 32) notes, "migration has transformed cities into veritable cultural mosaics, where identities mix and mingle".

Map (01) : Migration flows

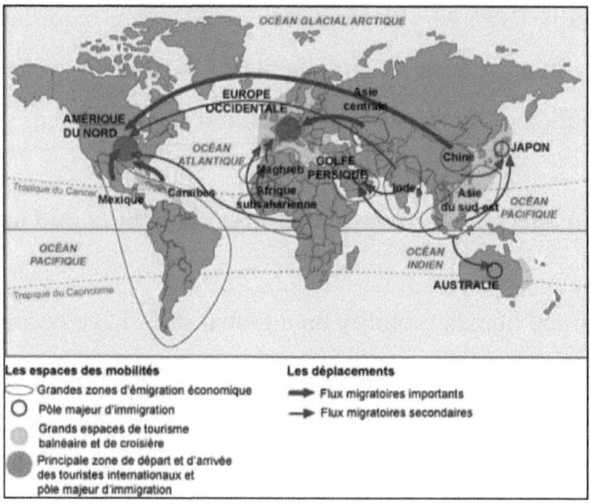

Source: https://www.maxicours.com/se/cours/les-migrations-et-le- tourisme-dans-le-monde/

Nevertheless, these migration-induced urban transformations also raise issues in terms of social cohesion, managing diversity and equitable access to public services. As Naïk Miret (2018, p. 112) points out, "policies for the integration and inclusion of migrants in cities are crucial to avoid social tensions and promote the harmonious coexistence of different communities". Global human mobility represents both a challenge and an opportunity for cities, requiring appropriate planning and development strategies to capitalise on the cultural and economic wealth of migration, while promoting social cohesion and equitable access to urban resources.

2. SPATIAL DUALITY IN CITIES: CONTRASTS BETWEEN PAST AND PRESENT

As a living, evolving space, the city reflects the social, economic and cultural transformations that shape our society. Over time, it undergoes metamorphosis, carrying with it the vestiges of the past and the aspirations of the present. This spatial duality, marked by striking contrasts between historical legacies and contemporary realities, is an intrinsic feature of the urban environment. It is precisely this tension between past and present that gives cities their richness and complexity, providing fertile ground for exploring the dynamics underlying their constant evolution.

2.1 The old town: deep historical roots and a rich urban heritage

Ancient towns are the repositories of a priceless historical heritage, living witnesses to bygone eras and the civilisations that shaped them. They bear the indelible marks of time, engraved in their streets. their majestic monuments and iconic architecture. These urban treasures, the fruit of long evolution, are the deep roots that anchor cities in their past, giving them a unique identity and an unrivalled wealth of heritage. To explore these historic districts is to plunge into a journey through the ages, where every stone tells a story, and every building bears witness to human ingenuity and creativity over the centuries.

2.1.1 The traditional urban fabric and heritage architecture: values and symbols of identity

Old towns have a rich historical and cultural heritage, embodied in their traditional urban fabric and heritage architecture. These tangible and intangible elements are markers of collective identity, conveying values and symbols that are deeply rooted in the memory of communities. According to Françoise Choay (1992, p. 159), "the ancient urban heritage is the irreplaceable witness of a particular culture, it is the bearer of the aesthetic and spiritual values that underpin the identity of each city". This view underlines the essential role of built heritage in preserving the authenticity and uniqueness of places.The traditional urban fabric, made up of narrow, winding streets, lively squares and old buildings, reflects the historical development of the town and bears witness to past lifestyles. As Pierre Merlin (2010, p. 231) points out, "traditional urban morphology is the result of a long stratification of forms and practices, shaped by geographical constraints, construction techniques and cultural traditions". Heritage architecture is of particular symbolic importance. Monuments, religious buildings, palaces and old mansions embody the collective memory and artistic aspirations of an era. As Jean- Pierre Poulain (2009, p. 87) notes, "heritage buildings are tangible witnesses to history, conveying cultural, social and aesthetic meanings".Beyond their material dimension, these heritage elements are invested with intangible values, such as emotional attachment, a sense of belonging and pride in identity. Françoise Navez-Bouchanine (2002, p. 42) points out that "the ancient urban heritage is a powerful vector of collective identification, reinforcing the link between inhabitants and their built environment". However, the preservation of this urban heritage faces many challenges, such as property pressure, the deterioration of buildings and the constraints associated with urban development. As Jean-Claude Driant and Meriem Chabou (2015, p. 127) point out, "safeguarding ancient urban heritage requires an integrated approach, reconciling the imperatives of conservation with the needs of sustainable urban development". Thus, the traditional urban fabric and heritage architecture are essential elements of the identity and collective memory of old towns, requiring sustained efforts to preserve and enhance them in order to pass on this heritage to future generations.

2.1.2 Social and economic challenges: poverty, unemployment and marginalisation

Despite their rich historical and cultural heritage, older towns often face major social and economic challenges. economic issues, such as poverty, unemployment and the marginalisation of certain populations. These issues are closely linked to the dynamics of urban transformation and the challenges of sustainable development. Urban poverty is a persistent reality in many older cities, where pockets of precariousness and social exclusion coexist. As François Ascher (2008, p. 95) points out, "historic city centres are often home to disadvantaged populations faced with substandard housing conditions, limited access to basic services and restricted economic opportunities".

This situation is exacerbated by high unemployment rates, particularly among young people and the low-skilled. According to Yves Jouffe and Jean-Marie Zembri (2013, p. 211), "economic and industrial change has led to the deindustrialisation of older urban centres, exacerbating the problems of unemployment and job insecurity".Social and spatial marginalisation is another major issue in older cities. Certain historic districts are often stigmatised and perceived as areas of relegation, reinforcing the processes of segregation and exclusion. As Jacques Donzelot (2006, p. 67) points out, "disadvantaged older neighbourhoods are often marked by a concentration of social and economic difficulties, creating a vicious circle of impoverishment and urban decay".

These social and economic challenges have a direct impact on the quality of life of residents and can threaten the preservation of urban heritage. As Dominique Lorrain (2011, p. 145) notes, "the deterioration in living conditions and the lack of economic prospects can lead to an exodus of This is leading to population growth and the gradual abandonment of older buildings, threatening the long-term survival of the urban heritage". To meet these challenges, many older towns have put in place strategies for economic regeneration and urban renewal, aimed at improving the living conditions of their residents and enhancing their heritage. However, these initiatives must be carried out in a balanced way, avoiding excessive gentrification and involving local people in defining development priorities (Françoise Navez-Bouchanine, 2002, p. 89).The social and economic challenges facing older towns require a comprehensive and integrated approach, combining action to promote employment, housing, education and social cohesion, while preserving the cultural and heritage identity of these unique urban areas.

2.2 The contemporary city: Manifestations of the modern era and growing impacts

As a counterpoint to the historic districts, the contemporary city is emerging as a tangible manifestation of the modern era. Shaped by the imperatives of progress, innovation and efficiency, it is characterised by bold construction, state-of-the-art infrastructure and forward-looking architecture.Towering skyscrapers, flowing arterial roads and redesigned public spaces bear witness to man's growing imprint on his urban environment. This new reality, driven by technological advances and emerging societal trends, is influencing lifestyles, travel patterns and social interactions within the city. The impact of modernity can also be seen in the challenges we face environmental issues, the challenges of sustainability and the dynamics of gentrification that are inexorably transforming the contemporary urban landscape.

2.2.1 Modern buildings and state-of-the-art infrastructure: Emblems of modernity and progress

Contemporary cities are characterised by their ever-changing urban landscape, marked by the emergence of modern buildings and state-of-the-art infrastructure, emblematic of modernity and progress. These architectural and technical developments reflect the ambitions of cities to be part of a dynamic of economic growth and innovation.According to Cynthia Ghorra-Gobin (2015, p.

172), "skyscrapers, office towers and high-tech real estate complexes have become the new symbols of the economic power and international influence of cities". These modern buildings, often designed by renowned architects, embody the desire of cities to attract investment and position themselves as key players in globalisation. Beyond their symbolic dimension, these buildings also meet functional needs linked to the growth of economic activities and changes in urban lifestyles. As Dominique Lorrain (2011, p. 203) points out, "the latest generation of office buildings offer flexible and innovative workspaces, adapted to the requirements of modern businesses and mobile workers". At the same time, today's cities are equipped with state-of-the-art transport and communications infrastructures, such as efficient public transport networks, international airports and high-speed fibre-optic networks. According to Marie-Christine Fourny and Marie-Flore Mattei (2010, p. 114), "these Infrastructure is essential to facilitate the mobility of people, goods and information, and to strengthen the competitiveness and attractiveness of cities". However, these urban developments are not exempt from criticism and questioning. Some criticise the standardisation of urban landscapes and the loss of local identity caused by the proliferation of standardised buildings. As François Ascher (2008, p. 145) points out, "the race for architectural modernity can lead to a certain trivialisation of urban forms, erasing the cultural and historical specificities of towns".What's more, these ambitious projects often raise environmental and social issues, such as the impact on natural resources, gentrification and population displacement. According to Jacques Lévy (2008, p. 198), "major urban projects must be conceived from a sustainable development perspective, taking into account their impact on the environment, social mix and quality of life for residents". Modern buildings and state-of-the-art infrastructures, while representing symbols of modernity and progress, raise crucial questions about their harmonious integration into the existing urban fabric, their environmental sustainability and their ability to meet the needs of the entire urban population.

2.2.2Social and cultural impacts: Changes in lifestyles and societal values

As well as physical and architectural transformations, the advent of contemporary cities has brought profound changes in lifestyles and societal values. These social and cultural impacts reflect the profound changes that have taken place in our society. induced by globalisation and the emergence of new urban dynamics.According to Cynthia Ghorra-Gobin (2015, p. 207), "contemporary cities have become spaces of consumption and leisure, shaping new aspirations and lifestyles". This evolution can be seen in the proliferation of shopping centres, entertainment complexes and leisure areas, responding to the expectations of an urban population in search of new modes of consumption and entertainment.

At the same time, economic and technological change has given rise to new ways of organising work and managing time. As François Ascher (2008, p. 175) points out, "contemporary cities are marked by an acceleration in the pace of life, greater flexibility in working hours and a fragmentation of social time".

These changes have had an impact on social values and behaviour, calling into question traditional models of urban life. According to Jacques Lévy (2008, p. 231), "contemporary cities are the scene of a growing individualisation of lifestyles, a questioning of traditional family structures and a diversification of forms of living".

In this context, urban public spaces have undergone a redefinition of their uses and meanings. As Dominique Lorrain (2011, p. 265) notes, "squares, parks and streets have become places for encounters, cultural expression and social events, reflecting the diversity of urban identities and practices".

However, these social and cultural transformations are not without their tensions and challenges. Some authors emphasise the risks of social fragmentation, loss of identity and cultural uprooting engendered by these rapid changes. As Françoise Navez-Bouchanine (2002, p. 124) points out, "contemporary cities must rise to the challenge of preserving their identity and social cohesion in the face of the homogenising forces of globalisation". The social and cultural impacts of contemporary cities bear witness to the complexity of today's urban dynamics, requiring in-depth reflection on how to reconcile individual and collective aspirations and preserve local identities while embracing the changes brought about by globalisation.

3. URBAN GLOBALISATION AND ITS IMPACT ON SPATIAL DUALITY: INTERSECTING OPPORTUNITIES AND CHALLENGES

Globalisation, an unavoidable phenomenon of our era, is having a profound influence on the spatial duality that characterises contemporary cities. On the one hand, it opens up new opportunities by facilitating cultural, economic and social exchanges on a planetary scale. Cities are becoming cosmopolitan crossroads, where diverse influences mix and mingle, enriching their urban fabric.On the other hand, globalisation raises major challenges, particularly in terms of integration, social cohesion and the preservation of local identities in the face of homogenisation. This tension between openness and rootedness can be seen in urban planning, where standardised modern spaces coexist with unique heritage districts.Urban globalisation highlights the complex issues involved in the harmonious coexistence of different temporalities and multiple cultural facets within a single city.

3.1 The growing challenges facing older cities in the context of globalisation

Old towns and cities, with their centuries-old heritage, face major challenges in today's globalised world. Firstly, demographic pressure and the influx of people from diverse backgrounds are putting a strain on historic infrastructures, which are often ill-suited to contemporary needs. Secondly, preserving the architectural authenticity and integrity of older districts is a constant challenge, in the face of the forces of modernisation and urban development. Thirdly, the balance between heritage enhancement and sustainable economic development is a delicate one, requiring careful management of resources and investment. Finally, rampant gentrification threatens the socio-cultural identity of these neighbourhoods, potentially leading to a loss of diversity and an erosion of the traditional social fabric. Faced with these complex challenges, older cities need to rethink their conservation, revitalisation and integration strategies, in order to preserve their rich heritage while adapting to the realities of globalisation.

3.1.1 Marginalisation and social and economic exclusion: risks of loss of identity and heritage

In the context of urban globalisation, older cities are facing major challenges linked to the marginalisation and social and economic exclusion of certain populations. These phenomena represent a direct threat to the preservation of the identity and heritage of these historic urban areas.According to François Ascher (2008, p. 211), "the processes of spatial relegation and socio-economic segregation have intensified in many older cities, leading to increased fragmentation of urban space". This dynamic has often led to the formation of deprived neighbourhoods, concentrating precarious populations facing difficulties in accessing housing, employment and essential services. This marginalisation has a profound impact on the social and cultural fabric of older cities. As Dominique Lorrain (2011, p. 287) points out, "socio-spatial exclusion generates a feeling of alienation and rootlessness among marginalised populations, threatening their sense of belonging and their attachment to the urban heritage".Indeed, when local populations are faced with precarious living conditions and a lack of opportunities, their ability to value and pass on their cultural and heritage heritage is seriously compromised. As Françoise Navez-Bouchanine (2002, p. 157) notes, "poverty and exclusion can lead to a gradual loss of traditional know-how, cultural practices and lifestyles associated with the ancient urban heritage".In addition, economic marginalisation can lead to increased property pressure on older neighbourhoods, threatening their physical integrity. According to Jacques Donzelot (2006, p. 98), "property speculation and poorly managed urban renewal can lead to the destruction or deterioration of heritage buildings, erasing traces of the history and identity of the area". Faced with these challenges, many cities have put in place strategies to revitalise and regenerate older neighbourhoods, with the aim of combating marginalisation and preserving the city's heritage. urban heritage. However, as Jean-Claude Driant and Meriem Chabou (2015, p. 174) point out, "these interventions must be carried out in an inclusive manner, involving local populations and taking care to avoid excessive gentrification processes that could lead to the uprooting of historic communities". The fight against marginalisation and social and economic exclusion is therefore crucial to guaranteeing the continued existence of the identity and heritage of old towns, ensuring that this cultural wealth is passed on to future generations and preserving the deep-rooted link between the local population and their built environment.

3.1.2 The erosion of identity and urban heritage: Tensions between preservation and development

In the context of urban globalisation, old towns face a major challenge:preserving their identity and cultural heritage. cultural heritage while respond to imperatives of economic development and modernisation. This tension between preserving the past and adapting to new contemporary realities is a major source of concern for many specialists in urban planning and urban sociology. According to Marie-Claude Smouts (2007, p. 23), "old towns are facing a real erosion of their identity and built heritage, under the pressure of economic forces and the logic of urban development". These forces, fuelled by globalisation, can lead to the destruction or radical transformation of radical transformation of sites historic sites, of heritage buildings and traditional neighbourhoods in favour of new housing developments, modern infrastructure or more profitable economic activities. This erosion of urban identity raises a number of cultural and social issues. As Françoise Choay (1992, p. 159) points out, "urban heritage is the bearer of identity, cultural and symbolic values that are essential for social cohesion and the sense of belonging of local populations". Its gradual disappearance can lead to a feeling of loss, uprooting and weakening of social ties within the communities concerned.Moreover, this tension between preservation and development highlights conflicts of interest between different urban players. On the one hand, property developers, investors and some local authorities advocate accelerated modernisation of cities, seeing heritage as a brake on economic development. On the other, heritage protection associations, local residents and some urban planning experts are campaigning for stricter preservation of historical features and identity (Rius, 2015, p. 87).

Faced with these challenges, several approaches have been proposed to reconcile heritage preservation and urban development. Some advocate the harmonious integration of heritage elements into urban renewal projects, while preserving their authenticity (Bourdin, 2008, p. 212). Others advocate a more radical approach, based on the enhancement and conversion of historic sites for sustainable cultural, tourist or economic purposes (Greffe, 2003, p. 97).

Be that as it may, this issue underlines the need for in-depth reflection on the management of urban identity in a context of accelerated globalisation, taking into account the multiple aspirations of the various stakeholders and by seeking a balance between preserving the past and adapting to the present.

3.2 The opportunities offered to contemporary cities in the context of urban globalisation

Urban globalisation brings its share of promising opportunities for today's cities. Firstly, openness to international flows of capital, talent and ideas stimulates innovation and economic growth. Cities are becoming magnets for multinational companies, skilled workers and foreign investment.

Secondly, the interconnection facilitated by new communication and transport technologies strengthens the links between the world's metropolises, promoting cultural exchanges and influence on the international stage. In addition, globalisation promotes greater diversity and mutual enrichment of urban cultures, nurturing the creativity and fulfilment of cosmopolitan communities.

Finally, shared environmental challenges encourage collaboration and the sharing of best practice in sustainable development between cities around the world. By seizing these opportunities, today's cities can position themselves as key players in globalisation and shape their future in a dynamic and competitive way.

3.2.1 Innovation and economic development: attracting investment and high added-value activities

In the context of urban globalisation, today's cities are faced with an unprecedented need to be economically competitive. To remain attractive and prosperous, they need to position themselves as hubs of innovation. innovation and development with high added value. In particular, this means being able to attract foreign investment and cutting-edge economic activities that create skilled jobs and generate sustainable growth.According to François Ascher (1995, p. 119), "today's cities must transform themselves into veritable platforms for innovation and wealth creation, relying on new technologies, research and development, and attracting the most competitive companies and talents". This economic development strategy, based on innovation, is seen as a guarantee of success in a globalised world where competition between metropolises is increasingly fierce. To achieve this, cities rely on various levers, such as the creation of competitiveness clusters, industrial clusters or free zones dedicated to high-tech activities. They also invest in cutting-edge infrastructure, universities and renowned research centres, in order to train a highly qualified workforce and attract innovative companies (Veltz, 2005, p. 237).

This approach is not without risks, as Saskia Sassen (2001, p. 87) points out: "The unbridled race for innovation and foreign investment can widen socio-economic inequalities within cities, and marginalise the most vulnerable populations". However, for many experts, it remains essential to ensure the competitiveness of today's cities in the age of globalisation.

Beyond the purely economic aspect, this strategy of development through innovation also aims to strengthen the attractiveness and international influence of our cities. By positioning themselves as centres of excellence in the following fields In addition to attracting investment, they are also seeking to attract talent, tourists and world-class events (Hamel and Brent Kato, 2010, p. 154). Urban globalisation offers contemporary cities the opportunity to reinvent themselves as nerve centres for innovation and wealth creation, provided they adopt the right strategies to attract investment and high value-added activities.

3.2.2 Attracting investment and tourism: boosting the region's image and international profile

In the context of urban globalisation, the appeal of today's cities goes beyond innovation and economic development. These cities are also seeking to boost their attractiveness to foreign direct investment (FDI) and international tourism, two key drivers of their growth and influence on a global scale.

According to Michel Lussault (2007, p. 112), "the world's major metropolises have entered into a genuine competition to attract capital flows, the head offices of multinational companies and foreign visitors". This involves urban marketing strategies aimed at promoting a distinctive brand image, highlighting the city's assets in terms of infrastructure, quality of life, and cultural and economic dynamism.

In terms of investment, cities seek to create a favourable environment for foreign companies by offering advantageous tax conditions, specialised business parks and a skilled workforce (Sassen, 2001, p. 167). They are also focusing on the construction of large, emblematic urban projects, such as office towers, shopping centres, and so on. and state-of-the-art facilities to project an image of modernity and prosperity.At the same time, attracting tourists is becoming a major issue for many cities. As Jean-Pierre Lozato- Giotart (2008, p. 87) points out, "urban tourism represents a considerable economic lever, in terms of direct and indirect financial spin-offs, as well as international influence". Cities are putting their heritage, their cultural offering and their flagship events on show, with the aim of attracting visitors from all over the world.However, this quest for attractiveness is not without risks. Some authors, such as Sharon Zukin (1995, p. 23), warn of the perverse effects of this 'museumification' of cities, which can lead to a standardisation of urban landscapes and a loss of authenticity. Others, such as Saskia Sassen (2001, p. 201), point to the social and spatial inequalities that can be created by the influx of investment and tourists to certain privileged districts. Despite these challenges, the race for attractiveness remains a priority for many of today's cities, which are keen to enhance their image and influence on the international stage against a backdrop of increasing competition between the world's major cities.

3.3 Tensions and contradictions between the two cities: Conflicts over resources and identities

Despite the benefits of globalisation, tensions and contradictions persist between the old and the new city. A major issue is the competition for limited resources, whether land, public budgets or private investment. While historic districts demand funds for their preservation, modern urban areas require massive investment in infrastructure and development projects. This dichotomy often leads to conflict and difficult trade-offs. Another point of friction is the question of identity. Defenders of heritage fight to preserve the authenticity of emblematic places, while promoters of modernity aspire to an urban identity that is resolutely forward-looking.This divergence of visions can lead to profound ideological and cultural clashes. In the face of these tensions, it is imperative to strike a delicate balance between preserving the past and developing the present, thereby promoting harmonious coexistence and fruitful synergy within the urban space.

3.3.1 The growing social and economic divide: Impact on social cohesion

While urban globalisation offers economic opportunities and international influence to today's cities, it also has perverse effects in terms of social inequalities and the fragmentation of social cohesion. The widening gap between the wealthiest sections of the population, who benefit from the effects of globalisation, and the underprivileged, who are marginalised by these changes, is a major challenge for many metropolises.

As Manuel Castells (1999, p. 165) points out, "the dualisation of the urban social structure is a striking feature of global cities, where a class of highly skilled workers rub shoulders with a growing fringe of poor or precarious workers". This social divide manifests itself through increased spatial segregation, with advantaged neighbourhoods benefiting from all the advantages of globalisation on the one hand, and disadvantaged areas with the most vulnerable populations on the other.This phenomenon of spatial and social dualisation gives rise to numerous tensions and risks for the cohesion of cities. As Jacques Donzelot (2004, p. 87) explains, "urban fragmentation and processes of territorial relegation encourage community withdrawal, the rise of identity-based discourse and the disintegration of social ties". Disadvantaged neighbourhoods thus become "urban ghettos", where the problems of unemployment, delinquency and marginalisation crystallise. Faced with these challenges, many researchers are calling for greater awareness and proactive public policies to combat these growing inequalities. According to François Ascher (1995, p. 211), "it is imperative to promote more balanced urban development, combining economic growth and redistribution of wealth, in order to preserve social cohesion and quality of life in cities". Among the avenues envisaged are the development of urban renewal and social mixing programmes, investment in infrastructure and public services in disadvantaged neighbourhoods, and the introduction of measures to support employment and entrepreneurship for the most vulnerable populations (Sachs, 2012, p. 145).Be that as it may, this issue underlines the need for in-depth reflection on the social impacts of urban globalisation and on ways of reconciling economic development with greater equity and better integration of all components of society.

3.3.2 Land and resource use conflicts: competing priorities and needs

The spatial duality that characterises many cities in the age of globalisation is not limited to social and economic dimensions alone. It also generates tensions and conflicts over the use of land and urban resources, highlighting the contradictions between the development priorities of ancient and contemporary cities. As David Harvey (2008, p. 32) points out, "the pressure on land from modern building projects and economic activities often conflicts with the need to preserve built heritage and historic districts". This competition for urban space reflects the divergence of interests between those involved in the old city, concerned with protecting their cultural heritage, and those in the contemporary city, focused on modernisation and economic development.

At the heart of these tensions is the crucial issue of managing land and property resources. In many cities, historic districts are coveted by property developers and investors, attracted by their potential for development and gentrification (Vivant, 2009, p. 67). This speculative pressure threatens not only the architectural heritage, but also the traditional social and economic fabric of these areas.

At the same time, the need for modern infrastructure and facilities, necessary for the development of contemporary cities, often conflicts with the preservation of natural spaces or established residential areas. As Henri Lefebvre (1968, p. 112) notes, "the capitalist logic of producing urban space tends to consume ever more land and environmental resources, to the detriment of the social and cultural uses of the city". Faced with these contradictory issues, many municipalities are trying to strike a balance between the different planning priorities. Some, as Françoise Choay (1992, p. 195) points out, are opting for a 'heritage' approach that aims to protect certain historic sectors while allowing measured development in other areas. Others, as Rem Koolhaas (1995, p. 67) explains, are exploring more radical urban models, in which the old city and the contemporary city coexist in juxtaposition, in a logic of 'urban collage'.

4. STRATEGIES FOR MANAGING SPATIAL DUALITY: TOWARDS SUSTAINABLE URBAN INTEGRATION

Faced with the challenges and opportunities presented by the coexistence of different eras within the urban space, new strategies are emerging to manage this spatial duality in a sustainable way. The aim is to achieve a harmonious integration of the different components of the city, while preserving their respective richness and authenticity.

This requires a comprehensive, cross-disciplinary approach, combining the imperatives of heritage conservation, sustainable economic development, social inclusion and environmental resilience. By rethinking development models, encouraging synergies between the old and the new, and involving all stakeholders, it is possible to turn challenges into opportunities.

The challenge is to create vibrant, dynamic and welcoming cities, where past and present come together in a fruitful balance, providing a framework in which present and future urban communities can flourish.

4.1 Integrated urban planning: A global vision for the future of cities

Integrated urban planning is emerging as an essential approach to meeting the challenges of spatial duality. This global, cross-disciplinary vision aims to harmonise the different components of the city, taking into account their interactions and interdependencies. It involves simultaneously addressing issues relating to historical heritage, economic development, social inclusion and environmental sustainability.

Such planning encourages the active participation of all stakeholders, from local authorities and citizens to experts and private investors. Thanks to this collaborative approach, it is possible to develop coherent strategies that enhance the unique assets of older districts while facilitating the harmonious integration of modern development projects.

The ultimate aim is to create functional urban spaces that are attractive and respectful of local identity, offering a quality living environment for present and future generations.

4.1.1 Preserving heritage and identity: integrating heritage features into urban planning

Faced with the tensions and risks of erosion of urban heritage in the context of globalisation, many experts are advocating an integrated approach to urban planning, aimed at preserving elements of identity and history while allowing cities to develop harmoniously. According to Pierre Merlin and Françoise Choay (2015, p. 243), "safeguarding the built and cultural heritage must be seen as an essential component of urban development sustainable". This approach involves identifying and protecting sites, buildings and districts with significant heritage value, while integrating them harmoniously into urban renewal and development projects.Such an approach requires close collaboration between urban planners, architects, art historians and heritage associations. As François Tomas (2004, p. 87) points out, "heritage preservation should not be seen as a brake on development, but as an opportunity to enhance the identity and uniqueness of the city, in a context of increased competition between metropolises". In practical terms, this can take the form of regulatory protection measures, such as the classification of certain buildings or districts, as well as financial and tax incentives to encourage the rehabilitation and enhancement of heritage (Gravari-Barbas, 2005, p. 112). Urban renewal projects must also be designed to respect the architectural integrity and historic character of the areas concerned. In addition to physical preservation, this integrated approach must also take into account the intangible dimensions of heritage, such as traditions, craft skills and local cultural practices. As Ismail Serageldin (1999, p. 23) notes, "urban heritage is not just a collection of old buildings, but also a way of life, a culture and a collective memory that must be passed on to future generations".

In short, integrating heritage into urban planning is an essential condition for reconciling the development of towns and cities with the preservation of their identity and heritage. their cultural richness, against a backdrop of accelerating globalisation.

4.1.2 Sustainable and inclusive development: taking account of environmental, social and economic dimensions

In order to meet the challenges posed by spatial duality in cities in an era of globalisation, it is essential to adopt an urban planning approach focused on sustainable and inclusive development. This holistic vision aims to reconcile the imperatives of economic growth with environmental protection and the promotion of social cohesion.

As François Ascher (2008, p. 67) points out, "sustainable urban development means radically rethinking the ways in which cities are produced, consumed and organised, in order to minimise their environmental impact while guaranteeing a better quality of life for all city dwellers".

In environmental terms, this involves strategies to reduce the ecological footprint of cities, by promoting public transport, energy-efficient buildings, sustainable management of natural resources (water, energy, waste) and the preservation of green spaces (Emelianoff, 2007, p. 32). The aim is to create cities that are more resilient in the face of the challenges of climate change and increasing scarcity of resources.

At a social level, sustainable development means combating inequality and urban fragmentation, by promoting better integration of disadvantaged populations and greater social diversity. This can take the form of affordable housing programmes, regeneration of deprived neighbourhoods, improved public services and access to employment for all (Sachs, 2012, p. 156). Finally, in economic terms, the aim is to encourage sustainable growth that respects the environment, by promoting activities with a low ecological impact, the circular economy and investment in green technologies (Camagni, 1998, p. 107). This holistic approach requires participative governance, involving all the players concerned (public authorities, businesses, civil society) in the development and implementation of urban policies. It must also take into account the specific local characteristics and differentiated needs of the population, to ensure real inclusion and ownership of projects by city dwellers. Although complex to implement, this vision of sustainable and inclusive urban development seems essential to meet the challenges posed by the spatial duality in the world's cities, by promoting economic growth that respects the environment and social cohesion.

4.2 Community participation and good governance: Foundations for successful planning

Successful urban planning, capable of meeting the challenges of spatial duality, rests on the solid foundations of community participation and good governance. The active involvement of citizens, from the earliest stages of the process, is crucial to ensuring that development projects meet the real needs and aspirations of local people. This inclusive approach enables diverse perspectives to be taken into account, cultural identities to be preserved and collective ownership of urban transformations to be fostered.Transparent, accountable governance based on democratic principles is essential to ensure the equitable management of resources and investments. It fosters mutual trust between the authorities, the private sector and civil society, creating a climate conducive to fruitful partnerships. By encouraging dialogue and synergy between all the players involved, community participation and good governance lay the foundations for sustainable urban planning, capable of harmoniously reconciling the demands of the present with the legacies of the past.

4.2.1 Involve all stakeholders: Representation of different groups and interests

To meet the challenges posed by spatial duality in cities in the age of globalisation, it is essential to adopt a participatory and inclusive approach to urban governance, involving all the relevant stakeholders. This approach aims to ensure that the various interest groups are fairly represented and that their respective aspirations are taken into account in the development of urban policies and projects.

As Patrick Le Galès (2003, p. 112) points out, "the city is a space for cohabitation and negotiation between a multitude of players with sometimes divergent interests: public authorities, businesses, associations, residents, etc. Effective governance means involving them all in the decision-making process". Effective governance means involving them all in the decision-making process".

This participatory approach has several advantages. Firstly, it allows the knowledge and expertise of each group to be integrated, thereby enriching the thinking on complex urban issues (Healey, 1997, p. 201). Secondly, it encourages ownership of the projects by the various stakeholders, reducing the risk of subsequent conflicts and disputes.However, implementing such participatory governance is not without its challenges. Care must be taken to ensure equitable representation of different groups, including the most vulnerable or marginalised populations, who are often under-represented in decision-making processes (Bacqué and Gauthier, 2011, p. 45). It is also crucial to put in place transparent and inclusive consultation mechanisms that allow for real dialogue and consideration of different points of view. The tools used include joint construction workshops, public consultations, neighbourhood committees and online citizen participation platforms. These mechanisms must be adapted to local realities and to the specific characteristics of the populations concerned (Sintomer et al., 2008, p. 87).Ultimately, involving all the stakeholders in urban governance appears to be an essential prerequisite for reconciling the various interests at stake and developing sustainable urban policies that are accepted by all, thereby promoting better integration of spatial duality in globalised cities.

4.2.2 Transparency and accountability: ensuring fairness and efficiency in decision-making

As well as involving the various stakeholders, effective and legitimate urban governance to meet the challenges of spatial duality must also be based on the fundamental principles of transparency and accountability. These principles are essential to guarantee fairness, responsibility and mutual trust in the decision-making process. According to Patrick Le Galès (2011, p. 57), "transparency and accountability are sine qua non conditions for the democratic and open management of urban affairs". This means making all relevant information on urban projects, decision-making processes and the use of public resources public and accessible.

Transparency thus enables citizens to monitor the actions of local authorities and the other players involved, reducing the risks of corruption, conflicts of interest or opaque decisions favouring certain groups to the detriment of others (Bingham et al., 2005, p. 548). It strengthens the credibility and legitimacy of the governance process in the eyes of the population. Accountability, for its part, implies that political and administrative leaders are held to account for their decisions and actions, and can be punished in the event of failings or abuses. As Yves Cabannes (2004, p. 32) points out, "accountability is a key element in empowering citizens and enabling them to have a real influence on the choices that affect them".

These principles can be implemented through various concrete measures, such as the regular publication of detailed reports on urban projects, the holding of public hearings, the establishment of complaints and appeal mechanisms, and the creation of independent monitoring and assessment bodies (Papadopoulos,

2007, p. 475).However, transparency and accountability are not easy objectives to achieve in a complex urban context, where multiple interests are at stake. They require a constant effort on the part of all stakeholders, as well as a framework and institutional framework guaranteeing their effective implementation (Piolatto and Lefebvre, 2016, p. 101). Ultimately, these principles appear to be essential conditions for equitable, effective urban governance that respects the different groups and interests involved, thereby making it possible to meet the challenges posed by spatial duality in cities in the age of globalisation.

4.3 Integrating past and present: a balanced approach to urban development

At the heart of efforts to manage the spatial duality of cities is the need to adopt a balanced approach to urban development, judiciously integrating the past and the present. This involves enhancing and preserving the rich historical heritage, while embracing the opportunities offered by modernity.

Older districts, which bear witness to the local identity, must be revitalised and restored with care, preserving their architectural authenticity and unique character. At the same time, contemporary development projects must blend harmoniously into the existing landscape, respecting the scale and character of the area.

Architectural innovation can coexist with heritage conservation, creating dynamic urban spaces where tradition and modernity are skilfully combined. This judicious integration requires in-depth consideration of uses, materials and styles, to ensure visual and functional coherence.

By embracing this balanced approach, cities can make the most of their historical and contemporary assets, offering a rich and stimulating living environment for all their residents.

4.3.1 Combining heritage and modernity: Making the most of everyone's potential

In the context of the spatial duality that characterises many cities in the age of globalisation, one of the major challenges is to reconcile the preservation of historical and cultural heritage with the imperatives of urban development and modernisation. Far from being incompatible, these two dimensions can in fact be mutually supportive, provided that an integrated approach is adopted to harness the respective potentials of heritage and modernity.

As Françoise Choay (1992, p. 159) points out, "urban heritage is not a hindrance to development, but an asset to be developed with a view to sustainable development and quality of life for city dwellers". Indeed, historic districts, emblematic buildings and local traditions are invaluable assets for the tourist appeal and cultural identity of cities.

However, as Rem Koolhaas (1995, p. 67) notes, "the contemporary city cannot be content with musing on its past, at the risk of becoming sclerotic". It is therefore essential to harmoniously combine heritage features with modern development projects that meet today's needs in terms of infrastructure, housing, public facilities and services.This combination of heritage and modernity can take many forms, from the respectful renovation of old buildings to their bold integration into contemporary urban complexes. Some cities, such as Barcelona and Bilbao, have managed to make the most of their historic heritage while at the same time welcoming avant-garde architectural creations, creating a stimulating synergy between the past and the future (Ingallina, 2008, p. 112).In addition to the physical aspect, this integrated approach must also take into account the intangible dimensions of heritage, such as traditional skills, cultural practices and local lifestyles. As recommended by UNESCO (2011, p. 23), "the safeguarding of living heritage must be accompanied by its enhancement and creative adaptation to contemporary realities". In this way, new economic or tourist activities can be developed around these intangible heritage elements, offering employment opportunities and sustainable development for local populations (Greffe, 2003, p. 87). Ultimately, combining heritage and modernity in urban projects is a promising way of meeting the challenges of spatial duality, by making the most of the riches of the past while embracing the potential of the present and future.

4.3.2 Harnessing technology and innovation: modern applications for sustainable solutions

In the quest for sustainable solutions to the challenges posed by the spatial duality of cities in the age of globalisation, the exploitation of modern technologies and innovation plays a crucial role. Far from being a threat to the preservation of urban heritage, these technological advances can, on the contrary, offer unprecedented opportunities to harmoniously combine historical heritage and contemporary development. As François Tomas (2004, p. 214) points out, "the new information and communication technologies (NICTs) are opening up new horizons for the promotion and interpretation of urban heritage". Thanks to augmented reality, 3D modelling and mobile applications, it is now possible to bring historic sites to life virtually, restore their original context and offer visitors immersive tours. These innovative digital tools make heritage more accessible and comprehensible to the general public, while preserving its material integrity (Gurriari, 2010, p. 67). They can also play a key role in the documentation and digital conservation of threatened architectural and urban remains. As well as enhancing our heritage, modern technologies also offer innovative solutions to the environmental and social challenges facing cities. For example, intelligent resource management systems (for water, energy and waste) make it possible to optimise their use and reduce the ecological footprint of urban areas (Caragliu et al., 2011, p. 54). Similarly, digital platforms for citizen participation or for reporting urban problems promote more open and inclusive governance, by directly involving residents in the management of their city (Neirotti et al., 2014, p. 32). However, the exploitation of these technologies must not be at the expense of authentic heritage and local identity. As Sharon Zukin (1995, p. 23) warns, "technology must not be a pretext for standardising and distorting the cultural specificities of cities". It is therefore essential to adopt a balanced approach, integrating technological innovations in a way that respects and adapted to local contexts, in consultation with the populations concerned.Ultimately, the judicious use of technology and innovation appears to be a promising lever for meeting the challenges of spatial duality in globalised cities, by making it possible to combine heritage preservation, sustainable development and improved quality of urban life.

CONCLUSION

The spatial duality that characterises many cities in the age of globalisation raises complex and multidimensional challenges. The sharp contrasts between ancient cities, rich in historical heritage, and contemporary cities, embodying modernity and economic development, give rise to profound social, environmental and cultural tensions. On the one hand, historic districts face the risk of losing their identity and heritage under the pressure of globalisation and urban development. On the other, modern areas concentrate investment, innovation and cutting-edge economic activities, but at the cost of widening socio-spatial inequalities.Faced with these contradictory challenges, it is essential to adopt a global and integrated approach to urban planning, taking into account the many dimensions of sustainable development. This means preserving the heritage and identity of older towns and cities, while enabling respectful and balanced modernisation that promotes innovation and international influence. This holistic vision requires the involvement of all stakeholders, with participative, transparent and accountable governance. It must also be based on a harmonious combination of heritage and modernity, taking advantage of the respective potentials of each, thanks in particular to the judicious use of modern technologies. Although complex to implement, this approach seems essential to meet the challenges posed by urban spatial duality and reconcile the sometimes contradictory aspirations of different social groups and territories. It offers the prospect of sustainable, inclusive urban development that respects the cultural identity of cities in a context of accelerating globalisation. Ultimately, managing this spatial duality appears to be a major challenge for the future of the world's metropolises, which are called upon to harmoniously combine their rich historical heritage with the imperatives of contemporary economic and social development.

BIBLIOGRAPHY

Ascher, F. (1995). Métapolis or the future of cities. Odile Jacob, France.

Ascher, F. (2008). Les nouveaux principes de l'urbanisme. Éditions de l'Aube, France.

Bacqué, M.-H. and Gauthier, M. (2011). "Participation, urban planning and urban studies". Participations, 1(1), France. p. 45.

Béal, V., Rousseau, M. and Vignon, S. (2018). World heritage city and sustainable development. Éditions Le Manuscrit, France. p. 11.

Bingham, L.B., Nabatchi, T. and O'Leary, R. (2005). "The New Governance: Practices and Processes for Stakeholder and Citizen.

Participation in the Work of Government". Public Administration Review, 65(5), U.S.A. p. 548.

Bourdin, A. (2008). Urbanisme et quartiers anciens. Presses universitaires de France, France. p. 212.

Cabannes, Y. (2004). "Municipal challenges for sustainable development". Cahiers des Amériques latines, 45(1), France. p. 32.

Camagni, R. (1998). "Sustainable Urban Development: Definition and Reasons for a Research Programme. International Journal of Environment and Pollution, 10(1), Netherlands. p. 107.

Caragliu, A., Del Bo, C. and Nijkamp, P. (2011). "Smart Cities in Europe". Journal of Urban Technology, 18(2), United Kingdom. p. 54.

Castells, M. (1998). La société en réseaux. Fayard, France. p. 379. Castells, M. (1999). The power of identity. Fayard, France. p. 165.

Choay, F. (1992). L'allégorie du patrimoine. Seuil, France. p. 159, 195.

Djament-Tran, G. and Fagnoni, E. (2015). "Metropolisations and heritage". Métropoles, 16, France. p. 8.

Donzelot, J. (2004). "La ville à trois vitesses : relégation, périurbanisation, gentrification". Esprit, 303(3/4), France. p. 87. Donzelot, J. (2006). Quand la ville se défait. Quelle politique face à la crise des banlieues? Seuil, France. p. 67, 98.

Driant, J.-C. and Chabou, M. (2015). Heritage and sustainable development. Une passion réciproque? L'Harmattan, France. p. 127, 174.

Emelianoff, C. (2007). "La ville durable: l'hypothèse d'un tournant urbanistique en Europe". L'Information géographique, 71(3), France. p. 32.

Fourny, M.-C. (2008). Ville mondiale. Presses universitaires de France, France. p. 12, 37.

Fourny, M.-C. and Mattei, M.-F. (2010). The urban economy and the new economy. Ellipses, France. p. 67, 114.

Ghorra-Gobin, C. (2015). Metropolisation in question. Armand Colin, France. p. 8, 42, 92, 125, 172, 207.

Gravari-Barbas, M. (2005). Habiter le patrimoine : Enjeux, approches, vécu. Presses universitaires de Rennes, France. p. 112. Gravari-Barbas, M. and Veschambre, V. (2005). "S'inscrire dans le patrimoine mondial". in Bourdin A., Hirschhorn M. et al, Les nouveaux défis de la géographie culturelle. Anthropos, France.

Greffe, X. (2003). La valorisation économique du patrimoine. La Documentation française, France. p. 87, 97.

Gurriari, L. (2010). "Valorisation virtuelle du patrimoine: nouveaux dispositifs, nouveaux comportements". Les Cahiers de la SFSIC, 4, France. p. 67.

Hamel, P. and Brent Kato, P. (2010). American entrepreneurial cities. Presses de l'Université du Québec, Canada. p. 154.

Harvey, D. (2008). "The Right to the City. New Left Review, 53, UK. p. 32.

Healey, P. (1997). Collaborative Planning: Shaping Places in Fragmented Societies. Macmillan, UK. p. 201.

Ingallina, P. (2008). Le projet urbain. Presses universitaires de France, France. p. 112.

Jouffe, Y. and Zembri, P. (2013). Geography of transport. De l'émergence aux mutations. Armand Colin, France. p. 211.

Koolhaas, R. (1995). S,M,L,XL. The Monacelli Press, United States.p. 67.

Le Galès, P. (2003). Le retour des villes européennes. Presses de Sciences Po, France. p. 112.

Le Galès, P. (2011). Le Retour des villes européennes (2nd edition). Presses de Sciences Po, France. p. 57.

Lefebvre, H. (1968). Le droit à la ville. Anthropos, France. p. 112. Lévy, J. (2008). L'invention du monde: une géographie de la mondialisation. Sciences Humaines, France. p. 68, 112, 198, 231. Lorrain, D. (2011). La main discrète. La Découverte, France. p. 23, 145, 203, 265, 287.

Lozato-Giotart, J.-P. (2008). Géographie du tourisme. Pearson Education, France. p. 87.

Lussault, M. (2007). L'homme spatial. Seuil, France. p. 112.

Ma Mung, E. (2009). "The Bronx district of New York in all its states". Géographie et cultures, 70, France. p. 32.

Merlin, P. (2010). L'urbanisme. Presses universitaires de France, France. p. 231.

Merlin, P. and Choay, F. (2015). Dictionnaire de l'urbanisme et de l'aménagement (4th edition). Presses universitaires de France, France. p. 243.

Miret, N. (2018). "Cities and migration: a complex equation". Hommes & Migrations, 1321(2), France. p. 112.

Navez-Bouchanine, F. (2002). La fragmentation en question. L'Harmattan, France. p. 42, 89, 124, 157.

Neirotti, P., De Marco, A., Cagliano, A.C., Mangano, G. and Scorrano, F. (2014). "Current trends in Smart City initiatives: Some stylised facts". Cities, 38, United Kingdom. p. 32.

Papadopoulos, Y. (2007). "Problems of 'governance' and 'deliberative democracy'". European Journal of Social Sciences, 45(136), Switzerland. p. 475.

TABLE OF CONTENTS

MIX
Papier aus verantwortungsvollen Quellen
Paper from responsible sources
FSC® C105338

Printed by Books on Demand GmbH, Norderstedt / Germany